DORLING KINDERSLEY *READERS*

PROFICIENT 4 READERS

SPACE STATION

ACCIDENT ON MIR

Written by Angela Royston

A Dorling Kindersley Book

Mike Foale
Michael Foale was born in Britain, but his mother is American. He moved to the USA because he wanted to become an astronaut.

NASA
These letters stand for National Aeronautics and Space Administration.

A great adventure

Jenna Foale gazed into the black nothingness of the night sky at Cape Canaveral, Florida. Her heart was thudding with excitement and fear. It was May 1997 and her father was about to be launched into space. A space shuttle would take him to Mir, the Russian space station that was circling around the Earth.

Jenna was five years old and her brother, Ian, was two. They did not know how dangerous the mission was, but they knew that they would miss their father while he was away.

Mike Foale was an astronaut. Like most American astronauts, he had spent only one or two weeks at a time in space before. This time, he was to live on the space station for four months. While he was there, he would be carrying out scientific experiments for the space organization NASA.

A Note to Parents and Teachers

Dorling Kindersley Readers is a compelling new reading programme for children, designed in conjunction with leading literacy experts, including Cliff Moon M.Ed., Honorary Fellow of the University of Reading. Cliff Moon has spent many years as a teacher and teacher educator specializing in reading and has written more than than 140 books for children and teachers. He reviews regularly for teachers' journals.

Beautiful illustrations and superb full-colour photographs combine with engaging, easy-to-read stories to offer a fresh approach to each subject in the series. Each *Dorling Kindersley Reader* is guaranteed to capture a child's interest while developing his or her reading skills, general knowledge, and love of reading.

The four levels of *Dorling Kindersley Readers* are aimed at different reading abilities, enabling you to choose the books that are exactly right for each child:

Level 1 – Beginning to read
Level 2 – Beginning to read alone
Level 3 – Reading alone
Level 4 – Proficient readers

The "normal" age at which a child begins to read can be anywhere from three to eight years old, so these levels are intended only as a guideline.

No matter which level you select, you can be sure that you are helping children learn to read, then read to learn!

Dorling Kindersley

LONDON, NEW YORK, SYDNEY, DELHI, PARIS,
MUNICH and JOHANNESBURG

Designed and edited by Bookwork

For Dorling Kindersley
Managing Editor Bridget Gibbs
Senior Art Editors Sarah Ponder, Clare Shedden
Senior DTP Designer Bridget Roseberry
Production Shivani Pandey
Picture Researcher Frances Vargo
Jacket Designer Yumiko Tahata
Illustrator Peter Dennis
Indexer Lynn Bresler
Consultant Air Commodore Colin Foale (father of Michael Foale)

Reading Consultant
Cliff Moon M.Ed.

Published in Great Britain by
Dorling Kindersley Limited
9 Henrietta Street
London WC2E 8PS

2 4 6 8 10 9 7 5 3 1

A CIP catalogue record for this book is available from the British Library

ISBN 0 7513 2934 7

Colour reproduction by Colourscan, Singapore
Printed and bound in China by L Rex

The publisher thanks the following for their kind permission to reproduce their photographs:
c=centre; t=top; b=below; l=left; r=right

Corbis: 9t (background), 11b, 39r; Reuters Newmedia Inc. 40l; Ted Streshinsky 42bl; **Corbis Sygma**: Jacques Tiziou 9b, R.P.G. 21b; **Eurospace Centre, Transinne, Belguim:** 27t; **John Frost Newspapers**: 'The Independent'/photo: AFP 16b; **Genesis**: 6-7, 12-13, 41b, 44t; **NASA**: 3, 4t, 7t, 9tl, 10t, 15t, 17, 20t, 22tl, 25t, b, 32, 33r, 35b, 36t, 42t, 45, 46b, 47t, b, 49; **Novosti, London**: 14tl; **Popperfoto/Reuters**: 13r, 16t; Viktor Korotayev 26l; NASA 36b; NASA/George Shelton 42-3; RTV 34; **Rex Features**: Brendan Beirne 31b; **Science Photo Library/NASA**: 2, 8t, 10bl, 19t, 20b, 28tl, 29, 41t, 44b; **Frank Spooner Pictures**: Gamma/Laski Diffusion 27b; **Topham Picturepoint**: 20br, 46; Associated Press 10br

Additional photography: Max Alexander 6t; Steve Gorton 4b, 12l, 15b, 35t; Alan Hills 19b; Gary Kevin 28b; Stephen Oliver 21t; Susanna Price 22bl; Science Museum/James Stevenson 14bl

see our complete catalogue at
www.dk.com

Contents

Mike Foale's mission would not be easy, but he had been preparing for it for a long time. He had trained with the Russian crew for 18 months and had learned to speak their language. At last the countdown to the launch had begun.

Growing seeds
One of Mike's experiments was to grow plants from seeds produced in space. The results would help future space station crews to grow their own food.

High-rise
The fuel tank of the space shuttle is as high as a five-storey building. It carries 2,345,736 litres of fuel – as much as ten jumbo jets.

The space shuttle stood ready on the launch pad. The spacecraft *Atlantis* was perched high above the ground, fixed to the back of a huge fuel tank. Two rockets were also fixed to the fuel tank. They would boost the spacecraft into the sky away from Earth.

Mike and the crew boarded *Atlantis* just after midnight on 16 May, and at 2:30 a.m. the hatch to the outside world was closed. Jenna and her family watched the launch pad from a safe distance.

The launch pad was brightly lit, but just after 4:00 a.m. the lights around *Atlantis* went out. Jenna gripped her mother's hand. The countdown continued: "Five, four, three, two, one ..."

A blazing flame burst from below the space shuttle as the engines roared into action.

"... Lift-off for the space shuttle *Atlantis*."

Countdown
The launch was controlled by a team of people and computers. Someone relayed the countdown through a loudspeaker to the people watching.

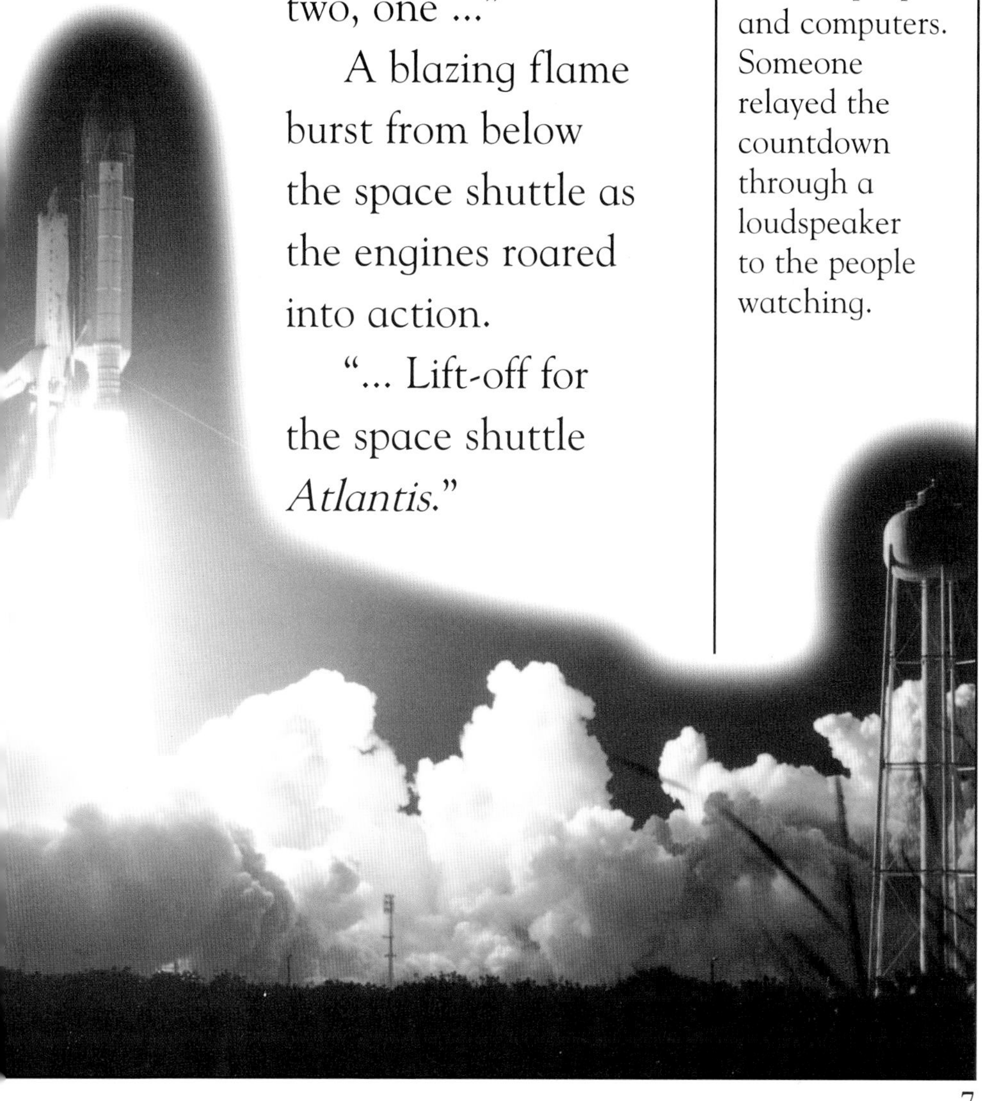

Recovery
The rocket boosters drift down on parachutes and are recovered from the sea by ships. They are then used again on another shuttle mission.

Atlantis lifted off into the night amid the roar of the engines. Mike watched the launch tower slip past and felt the juddering of the rocket boosters. There was no going back now! Two minutes later, the rockets had burned all their fuel and fell away from the shuttle. Then the juddering stopped.

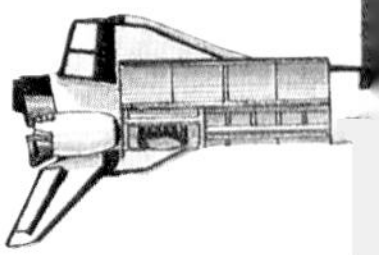

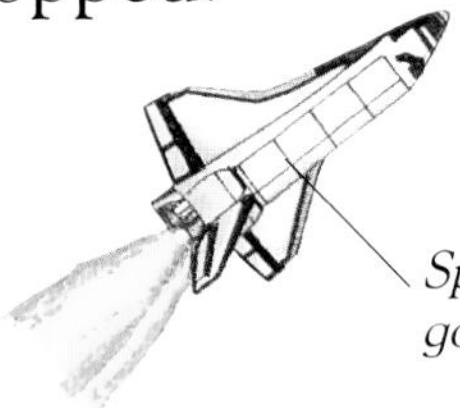

Spacecraft goes into orbit

Fuel tank falls away

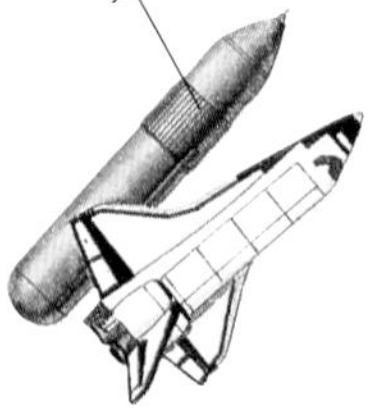

Rocket boosters fall away

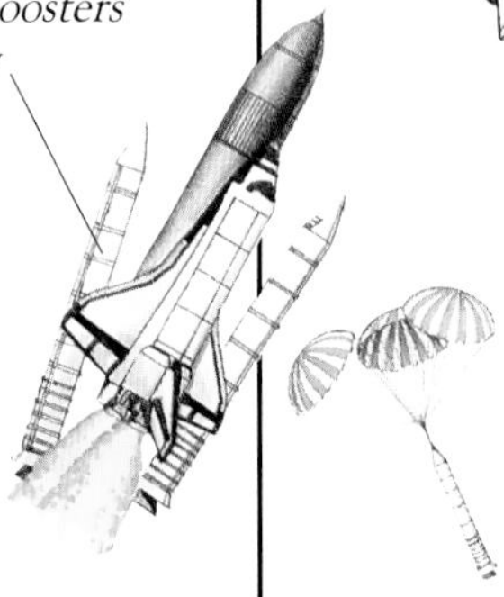

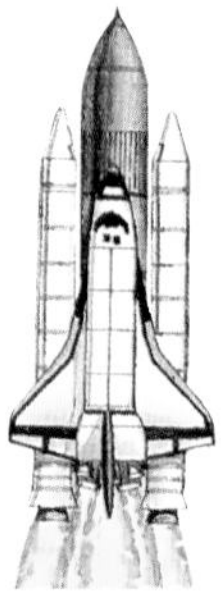

The watchers below saw the fiery trails of the falling rockets. For seven more minutes, the main engines took *Atlantis* higher and higher, until they had used up all the fuel in the tank. Then the huge empty tank also fell away, leaving the spacecraft to continue its journey on its own.

To everyone at Cape Canaveral, *Atlantis* was now just a speck of light in the night sky. The lift-off was completed! They all breathed a sigh of relief – except Jenna. "Daddy is turning into a star!" she sobbed. And indeed *Atlantis*, 320 kilometres above Earth, looked like a star in the sky.

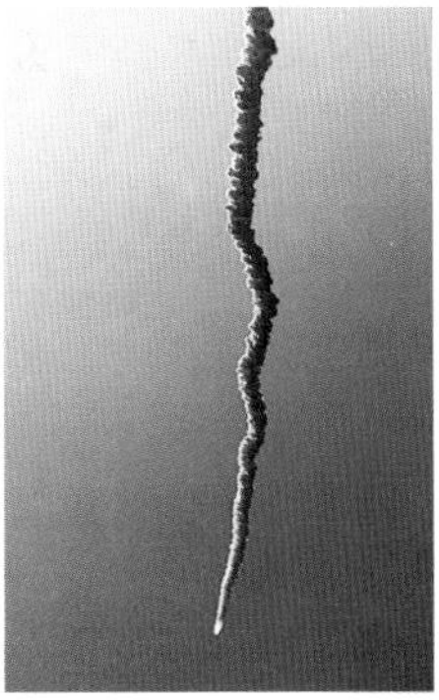

Burnout
The fuel tank burns up as it falls through the air.

Space sounds
Space is silent, but it is not quiet inside the shuttle. Air has to be pumped around, and this makes lots of noise.

Space record
In May 1997, 10 astronauts, including Mike Foale, set a record for the number of people in one orbiting craft.

Mike and the crew were relieved and delighted to be safely in space. The engines were turned off and the only noise was the hum of the spacecraft's computers and machines. *Atlantis* began to circle Earth once every 90 minutes.

The view was amazing. First the astronauts saw continents, oceans and clouds all bathed in beautiful, warm sunlight. Then, when the spacecraft travelled over the dark, night side of Earth, the astronauts could see the stars – many more than they could see from Earth, and much brighter.

A satellite image of Earth showing the dark and light side, or day and night

A view of the city of Tokyo, Japan, as seen from Atlantis.

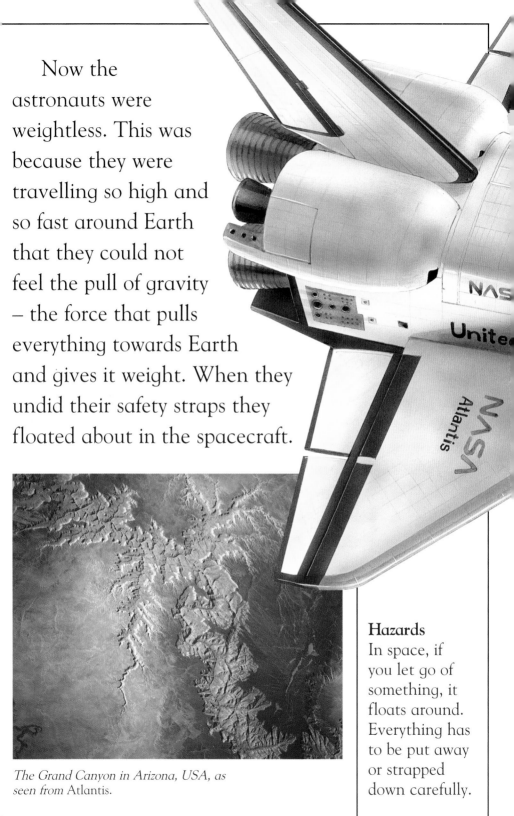

Now the astronauts were weightless. This was because they were travelling so high and so fast around Earth that they could not feel the pull of gravity – the force that pulls everything towards Earth and gives it weight. When they undid their safety straps they floated about in the spacecraft.

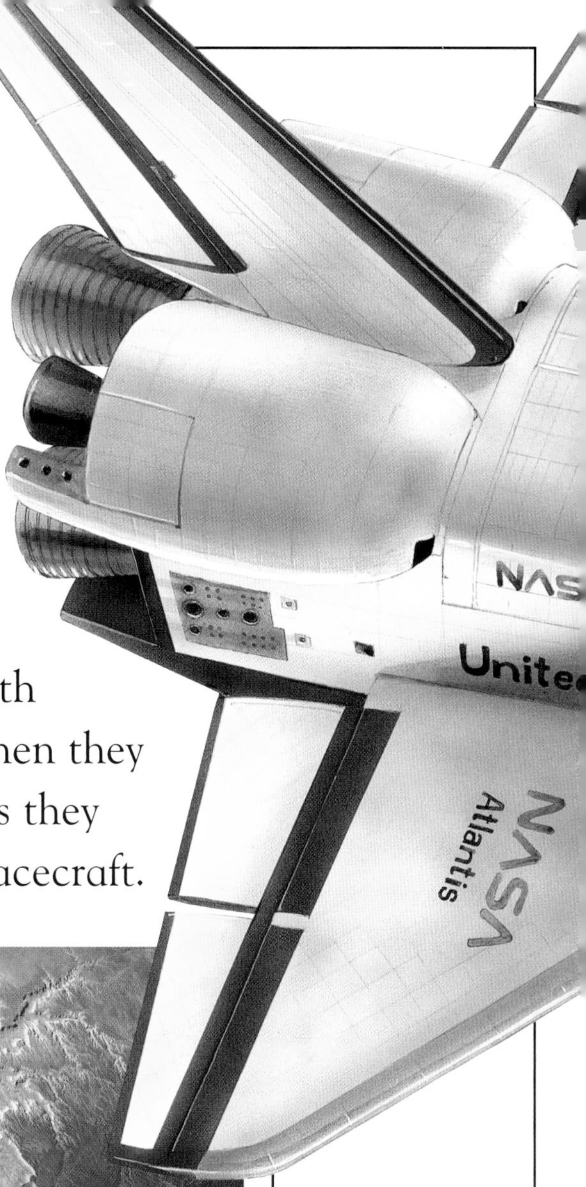

The Grand Canyon in Arizona, USA, as seen from Atlantis.

Hazards
In space, if you let go of something, it floats around. Everything has to be put away or strapped down carefully.

Cosmo or astro
In Russia, people who travel in space are called cosmonauts. In the USA, they are called astronauts.

Welcome gift
International space crews exchange gifts like these Russian sweets. Cosmonauts on Mir gave new astronauts a traditional gift of bread and salt as they came aboard.

Space call
NASA told Mike's family when Mir was overhead so that they could look out for it.

The crew began to prepare for the docking with Mir. The pilot slowly guided *Atlantis* closer and closer to Mir, until the spacecraft were lined up exactly. Carefully, he slid *Atlantis* into one of Mir's docking hatches. The hatch doors between Mir and *Atlantis* opened and Mike was greeted by the two Russian cosmonauts on Mir, Sasha and Vasily.

The mission and space station were controlled by Ground Control – a team of computers, scientists and experts in Russia. But Ground Control could communicate with Mir only when it was overhead.

Atlantis *docked with the Mir space station*

As soon as he could, Mike joined the Russian crew. He took off his NASA clothes and put on Russian cosmonaut clothes. The two crews unloaded supplies of water and equipment from *Atlantis.*

A few days later, *Atlantis* left Mir to return to Earth. Mike watched as it disappeared. It would be a long time before he saw his friends from NASA again.

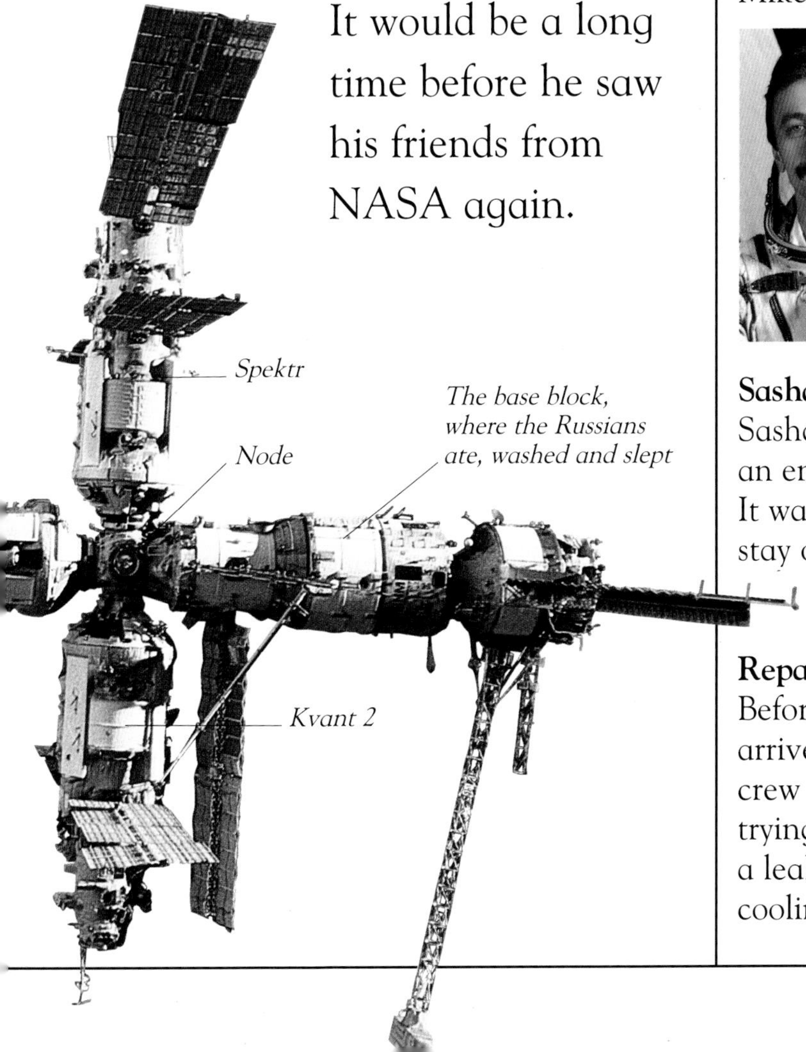

Vasily Tsibliev
Vasily was the commander on Mir when Mike arrived.

Sasha Lazutkin
Sasha was an engineer. It was his first stay on Mir.

Repairs
Before Mike arrived, Mir's crew had been trying to find a leak in the cooling system.

Bit by bit
Mir was built in space. Each module was attached one by one. This is Mir in 1986.

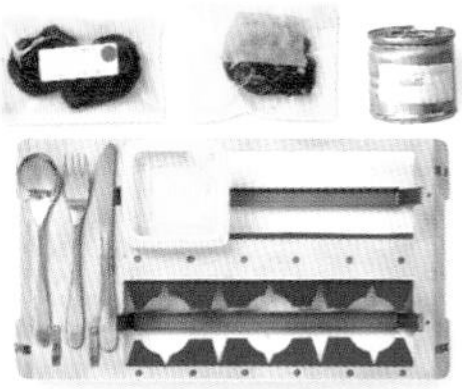

Space food
Space meals are pre-prepared on Earth, and the food is put in packs on a tray. The astronauts use fingers or cutlery to eat from one pack at a time.

The Mir space station was made up of several modules. Mike floated from one to another using his hands and feet to push himself along the walls. The largest module was the living area where the Russians ate, washed and slept. At one end was a round Node with hatches to other modules. One of these, Spektr, was where Mike slept.

Mike set up his experiments in the modules around the Node. These modules were already cluttered with machines and equipment. The module Kvant 2, for example, held the machines that made life in space possible. These included a machine called Elektron. It used waste water to make oxygen.

Sasha Lazutkin taking a photo inside Mir

All the machines used electricity. Solar panels on the outside of the station collected energy from sunlight and turned it into electricity. Machines called gyrodynes kept the station angled so that, as it turned, the solar panels always faced the Sun and the panels collected as much energy as possible.

Sun power
The electricity made in solar panels is stored in batteries. In space, it is used to provide heat and light and to power equipment.

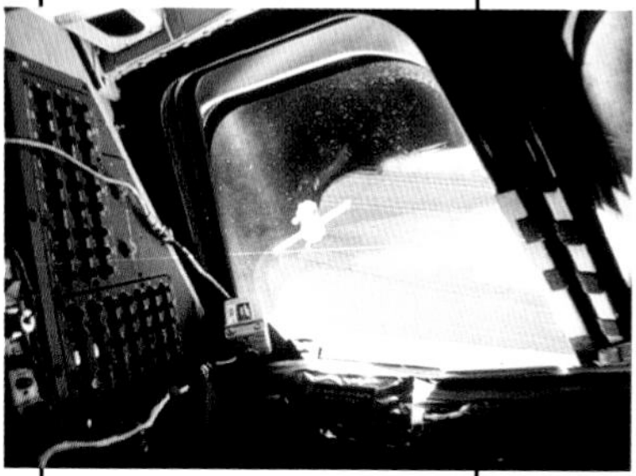

Progress
The docking test for Progress was dangerous. Vasily had to manoeuvre the craft manually from 6.5 km away from Mir to 50 metres away.

Disaster strikes

Things went well on Mir until 25 June. Then disaster struck. Vasily was practising a new way of docking the Progress supply ship. This craft brought new supplies from Russia and took away rubbish. It had no crew so Vasily was operating it by remote control.

Suddenly, Vasily realized that the supply ship was coming towards them too fast. He tried to make it brake, but it was too late. Progress smashed into the solar panels on Spektr and then hit the module itself.

Mike heard a thud and then a hiss. Air was escaping from the space station. The crew was in great danger. If all the air escaped, they would die in minutes. The air pressure alarms began to ring. They were so loud that the men could not hear each other speak.

THE INDEPENDENT
SOARING HOUSE PRICES, BUILD YOUR OWN
WHERE DAREN'T YOU BREASTFEED THESE DAYS?
FREE: ENJOY A THAI FOOD DISH UP TO THE VALUE OF £8
Mir, a disaster movie in the making

Bad news
On Earth, news of the crash on Mir soon appeared in newspapers and on television.

Sasha thought that Progress had made a hole in Spektr. He tried to shut the hatch so air would not escape from the whole station. But electricity cables were in the way.

Essential cables
Cables carried electricity from Spektr's solar panels to the rest of the space station.

Mir orbiting Earth after the accident

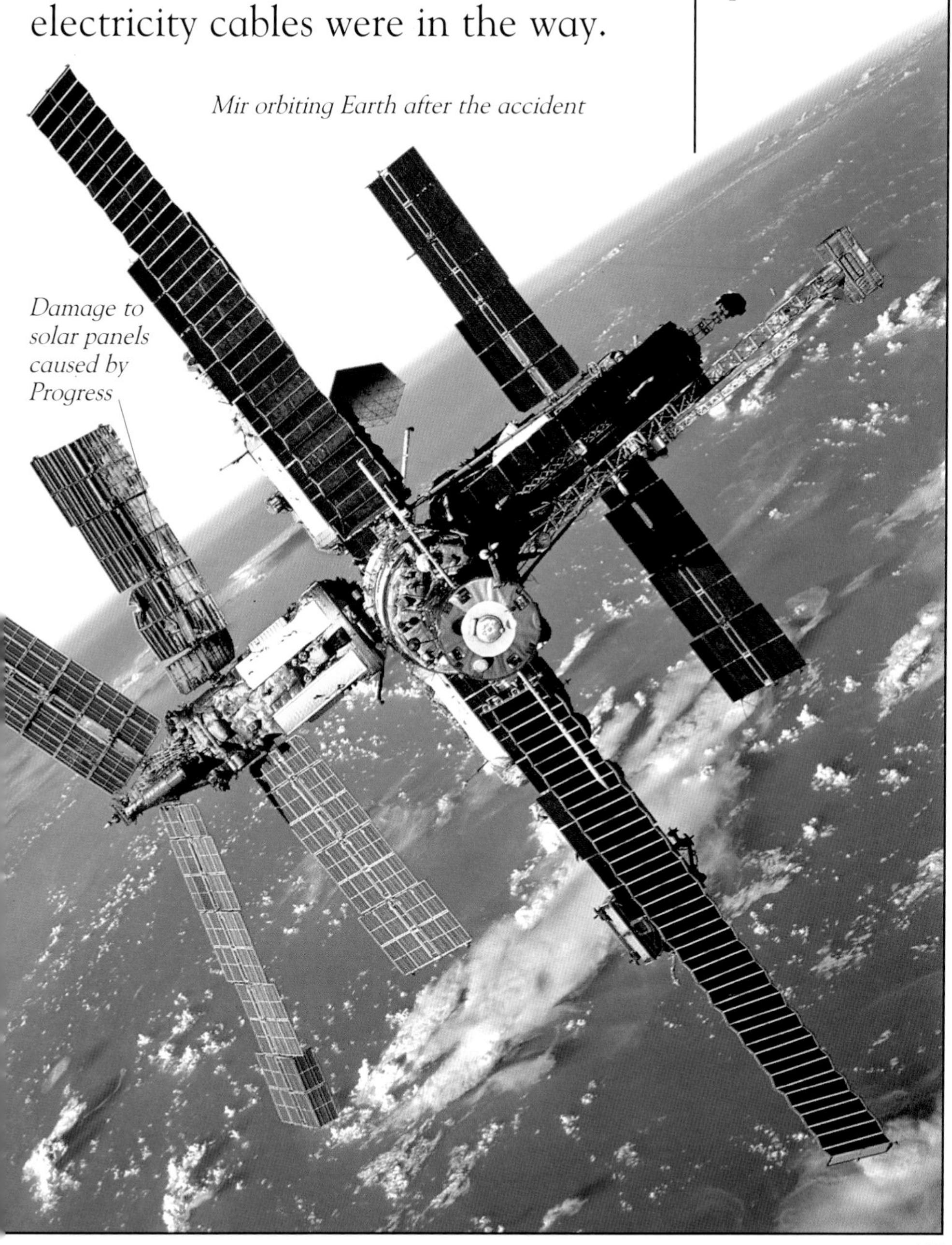

Damage to solar panels caused by Progress

Quickly, Mike came to help Sasha. They cut some of the cables and pulled out the plugs on the others. Then they slammed the hatch shut. They had stopped air escaping from the rest of the station, but they were not safe yet. Mir was spinning and tumbling in space, the solar panels had stopped making electricity, and the batteries were dangerously low.

Mike and Sasha switched off most of the batteries to save electricity. The lights went out and the space station became freezing cold. Then Mir moved into sunlight. The men could see again, but the station became boiling hot.

The situation was desperate but the cosmonauts had been trained to cope with emergencies. Elektron was not working, so they lit special candles that made oxygen instead of burning it so that they could breathe.

Freezing cold
On the dark side of Earth, the temperature in space is colder than in the Antarctic.

Boiling hot
When the Sun shone on Mir, the temperature quickly rose and became hotter than the hottest desert on Earth.

Space disasters
Space travel is dangerous. In 1969 there was an explosion on *Apollo 13* that put the engine out of action. Mission Control worked out how to get the spacecraft home safely.

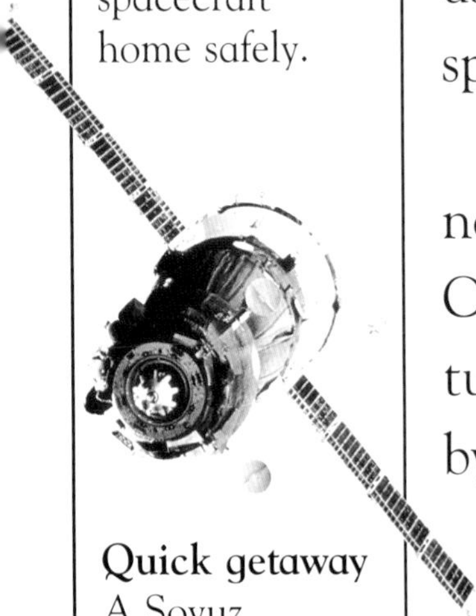

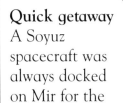

Quick getaway
A Soyuz spacecraft was always docked on Mir for the crew to use if they had to escape in an emergency.

The people at Ground Control tried to help the cosmonauts. The most important task was to stop the station from spinning. "What is your angle and speed?" they asked.

There was no instrument on board that could give the answer. Then Mike put his thumb to the window and estimated how fast it was moving compared to the stars. Now Ground Control was able to use Mir's engines to stop the space station spinning.

But the solar panels still did not face the Sun. Ground Control told Vasily to turn the space station by firing bursts of the thrusters in the docked spacecraft Soyuz – the craft that had brought Vasily and Sasha to Mir.

Mike again used the stars and the Sun to work out how long they should fire the thrusters. Easier said than done! The calculations were difficult and he was exhausted.

Vasily worked the engines. Mike wanted him to use more power but Vasily would not risk it. "We will need Soyuz to escape from Mir," he said. "If we use too much of its fuel, we will be stuck up here."

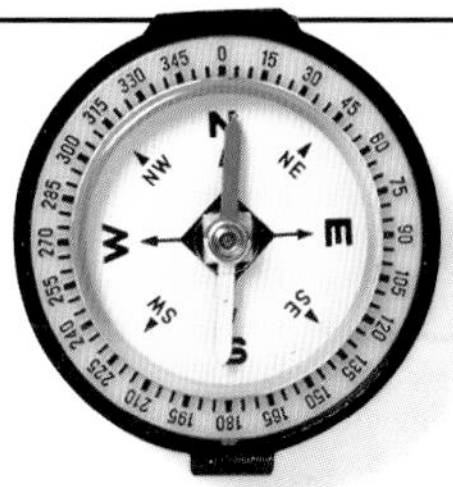

Star work
You can find out which way you are facing on Earth with a compass. It is much more difficult in space.

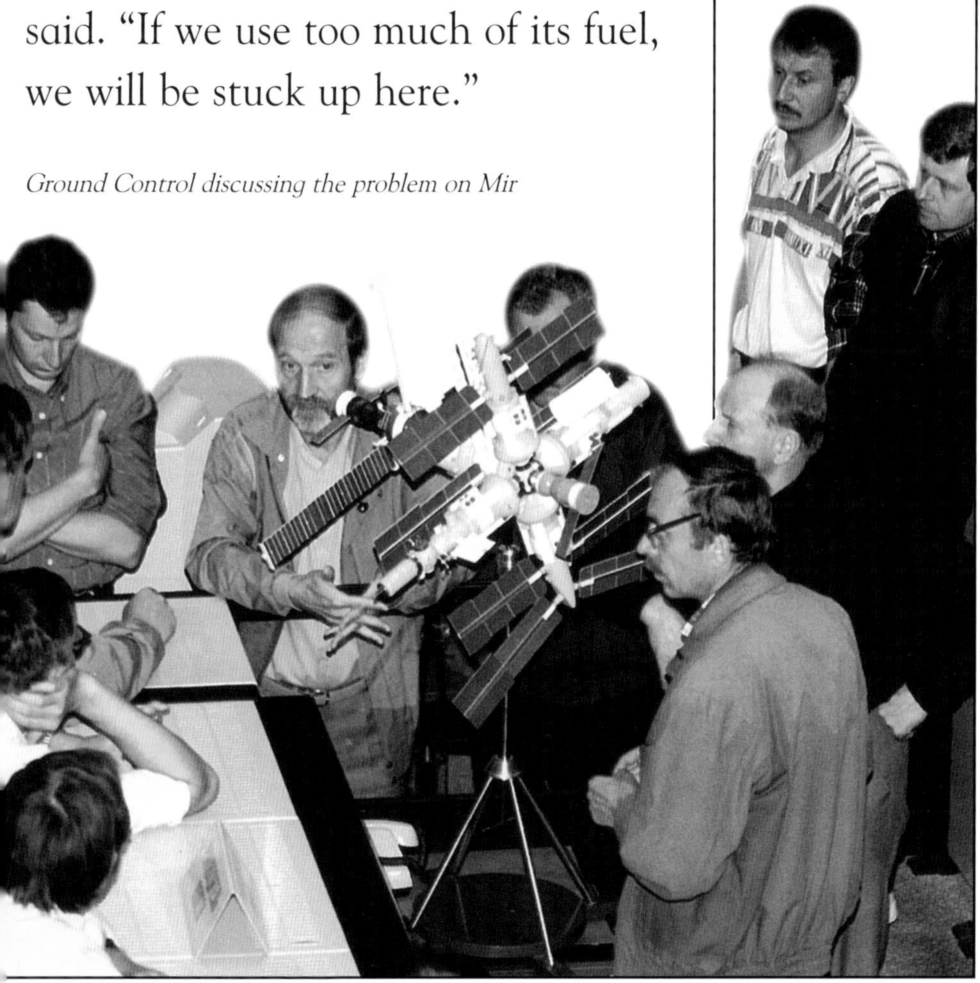

Ground Control discussing the problem on Mir

Day and night
Cosmonauts have to make their own day and night. When it's time for bed, they block the windows, turn off the lights, and sleep.

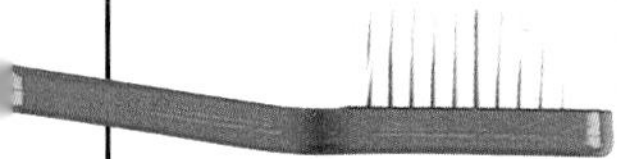

No toothbrush
When the hatch to Spektr was closed, Mike lost lots of his things including his sleeping bag, toothbrush and clothes.

Sasha persuaded Vasily that Mike was right. It took 24 hours to turn Mir into the correct position, but at last the solar panels faced the Sun again and the batteries started recharging. Now the crew could get some sleep.

Although the living area had light, heat and cooling again, there was not enough electricity to heat the other modules. Nearly half of Mir's electricity was made by the solar panels on Spektr, and the cables had been cut.

Ground Control had a plan. Vasily and Sasha would put on spacesuits, go into Spektr, and reconnect the cables. It would be like a spacewalk, or extravehicular activity (EVA) – a dangerous job. It was arranged for 17 July.

Vasily and Sasha started to practise. Vasily thought that people blamed him for the accident and he wanted to put it right. But he was exhausted and anxious, and his heart was not beating properly. Ground Control thought that he might have a heart attack.

Cold comfort
Once Spektr was closed, Mike slept with his head in the Node and his feet in Priroda, an unheated module. Vasily lent him a sleeping bag because Mike's was in Spektr.

Spacewalk
The first person to do an EVA was cosmonaut Alexei Leonov. His spacewalk took place in March 1965.

Ground Control asked Mike if he would go with Sasha to reconnect the cables instead of Vasily. Mike was happy to do it, but NASA was worried. How safe was Mir now?

Mike and Sasha went on preparing for the "spacewalk". All the hatches into the Node had to be sealed shut to stop air from escaping when Mike and Sasha opened the hatch to Spektr. But more than 100 electricity cables passed through the Node. All of these cables had to be unplugged before the hatches could be sealed. The crew practised unplugging the cables in the right order. This was important.

Two days before the "spacewalk", Sasha made a terrible mistake. He unplugged the cables that powered the gyrodynes, and Mir spun out of control again.

Once again Vasily and Mike had to use Soyuz's engines to turn Mir so that the solar panels faced the Sun. This time Vasily did exactly what Mike asked, and Mir was soon stable.

Working guest
Mike was a guest on Mir so he was not meant to help the Russians with their jobs. But after the accident, his help was needed more and more.

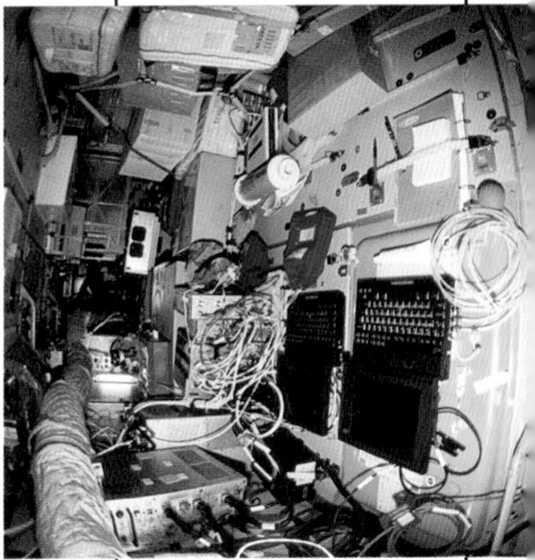

Jumble
So many cables passed through Mir, that it was difficult to tell which cable was attached to which machine.

Help arrives

Two new Russian cosmonauts were due to go to Mir soon, and Ground Control decided that they should do the spacewalk. This meant that life on board the space station would be difficult for several more weeks.

On Earth, Anatoli Solovyev and Pavel Vinogradov practised the spacewalk in a model of Spektr in a water tank – being underwater is the nearest thing on Earth to being weightless.

The cosmonauts launched into space on 5 August and arrived on Mir two days later. They took a new hatch for Spektr. It allowed cables to pass through it without letting air escape.

Anatoli (left)
The new Mir commander was Anatoli Solovyev. He was a very experienced cosmonaut.

Pavel (right)
It was Pavel Vinogradov's first time on Mir. He was the flight engineer, and he replaced Sasha, with whom he was good friends. His nickname was Pasha.

On 14 August, after six months on Mir, Vasily and Sasha left in their Soyuz spacecraft to return to Earth. Mike was sad to say goodbye to them – they had survived so much together.

"I hope that everything bad leaves with us," said Vasily. He thought that while he had been on Mir he had brought bad luck to the space station.

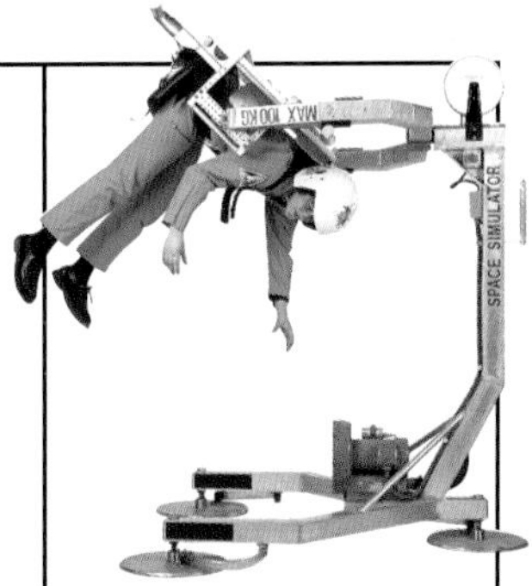

Weightlessness
Astronauts can prepare for the weightlessness of space. The feeling can be simulated in a "five degrees of freedom chair".

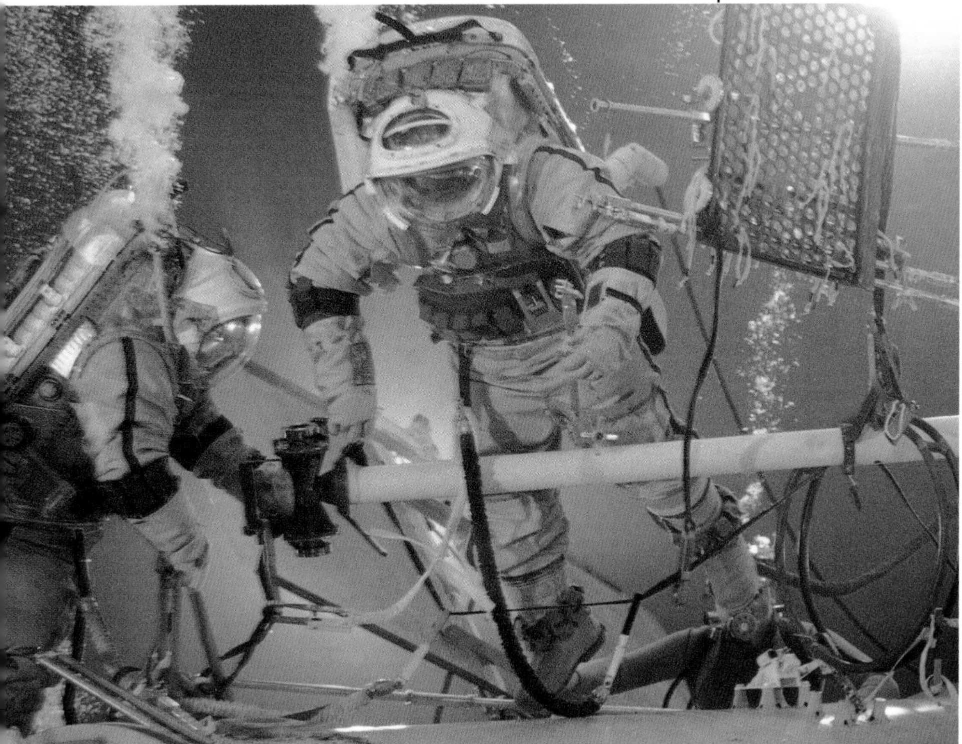

Top right: "Five degrees of freedom chair" Eurospace centre, Transinne, Belgium

Video link
Mike's video camera came in handy. His film of Spektr was transmitted to Ground Control.

Shutdown
The computer on Mir crashed just like the computers you use at your school or in your home sometimes do.

The new crew wasted no time. On 16 August, Anatoli, Pavel and Mike boarded the new Soyuz spacecraft, and Anatoli flew it around the outside of Mir so that they could see the damage.

Mike used a video camera to film the damage to Spektr. The film would be used to plan an EVA in which Anatoli and Mike could examine the damage more closely. This would take place after the spacewalk inside Spektr. They flew around Mir for 44 minutes before redocking.

The crew continued for many days to prepare for the internal spacewalk. Then bad luck struck again. Mir's main computer suddenly shut down.
The space station began spinning and tumbling out of control again.

Anatoli and Pavel were alarmed, but Mike was used to it by now! "Don't worry," he said. "I've done this before!"

Mir on TV
Ground Control often set up a video link so that people on Earth could see Mir on TV.

Outside view of the damage to Mir

Out of time
The spacesuits had enough oxygen to last for only six hours.

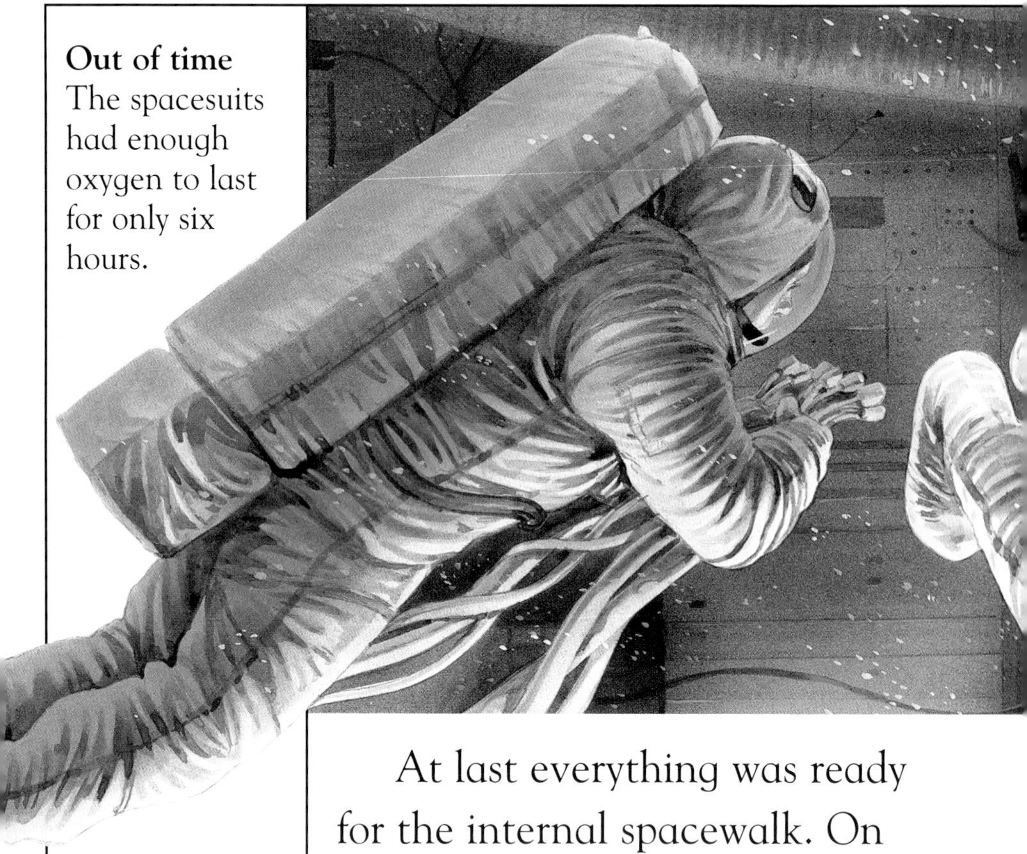

Clean crystals
When Anatoli and Pavel entered Spektr, there were lots of tiny white crystals floating about. They were crystals from Mike's shampoo, which had separated because there was no air pressure to keep it together.

At last everything was ready for the internal spacewalk. On 22 August Anatoli and Pavel put on their spacesuits. The spacesuits provided oxygen for the men to breathe and protected them from the extreme heat and cold of space. Mike floated into Soyuz, where he waited during the spacewalk. If anything went badly wrong, they would all escape in Soyuz. Mike could talk to the other two by radio, but he could not see them.

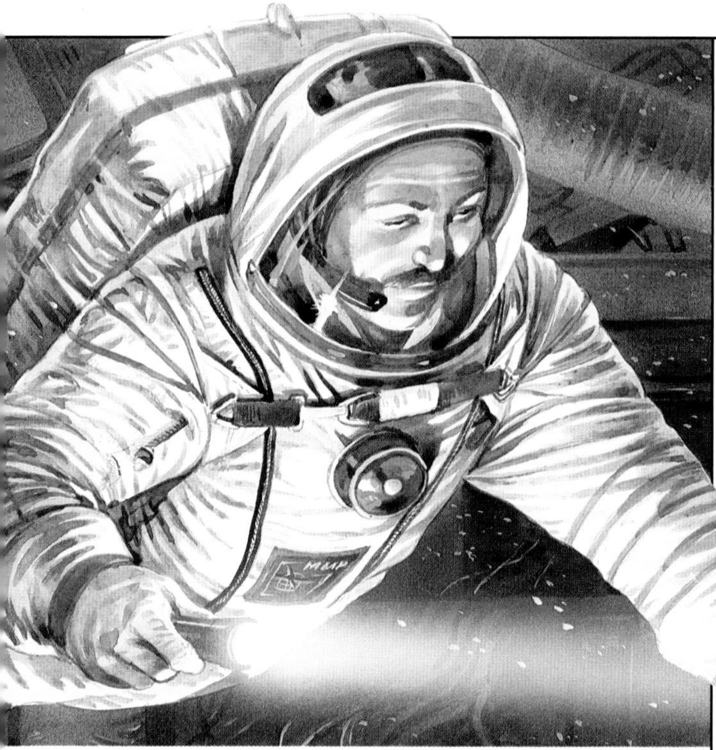

Spacesuits
The first spacesuits were each made for a particular astronaut, who wore it all the time. Now there are suits for different jobs, and they are reused.

Anatoli and Pavel sealed all the hatches to the Node. Then they removed all the air from the Node before opening the hatch to Spektr.

Using torches, Anatoli and Pavel looked around Spektr for the cables from the solar panels. Then they fitted the new hatch and began to reconnect the cables. It was hard work and it took them three hours. Before he left Spektr, Pavel gathered some of Mike's things, including a photo of his children.

Happy father
When Pavel told Mike he was bringing him the photo of his children, Mike said, "That will make me very happy."

Fixing things
Making repairs is part of an astronaut's job. An old station like Mir always had something that needed fixing using specially adapted space tools.

Jet packs
As well as tethers, US astronauts use manned manoeuvring units (MMUs). These have tiny gas jets that astronauts use to propel themselves around when in space.

It was one of the most difficult repairs that had ever been made in space. But the crew got their reward. By the next day, the batteries were fully charged and all the modules had light and heat again. Would the spacewalk outside Mir also go well?

Ground Control had planned the external spacewalk carefully. This was the plan: Mike would be tethered to Mir to stop him from floating off into space. Anatoli would be tethered to the end of a long boom. Mike would hold the boom while Anatoli tried to find the hole in Spektr. If he found it, he would fix it. Then the crew would be able to use the module again.

By 6 September they were ready. Mike and Anatoli put on their spacesuits and went into the airlock at the end of the module Kvant 2. This special hatch had two doors.

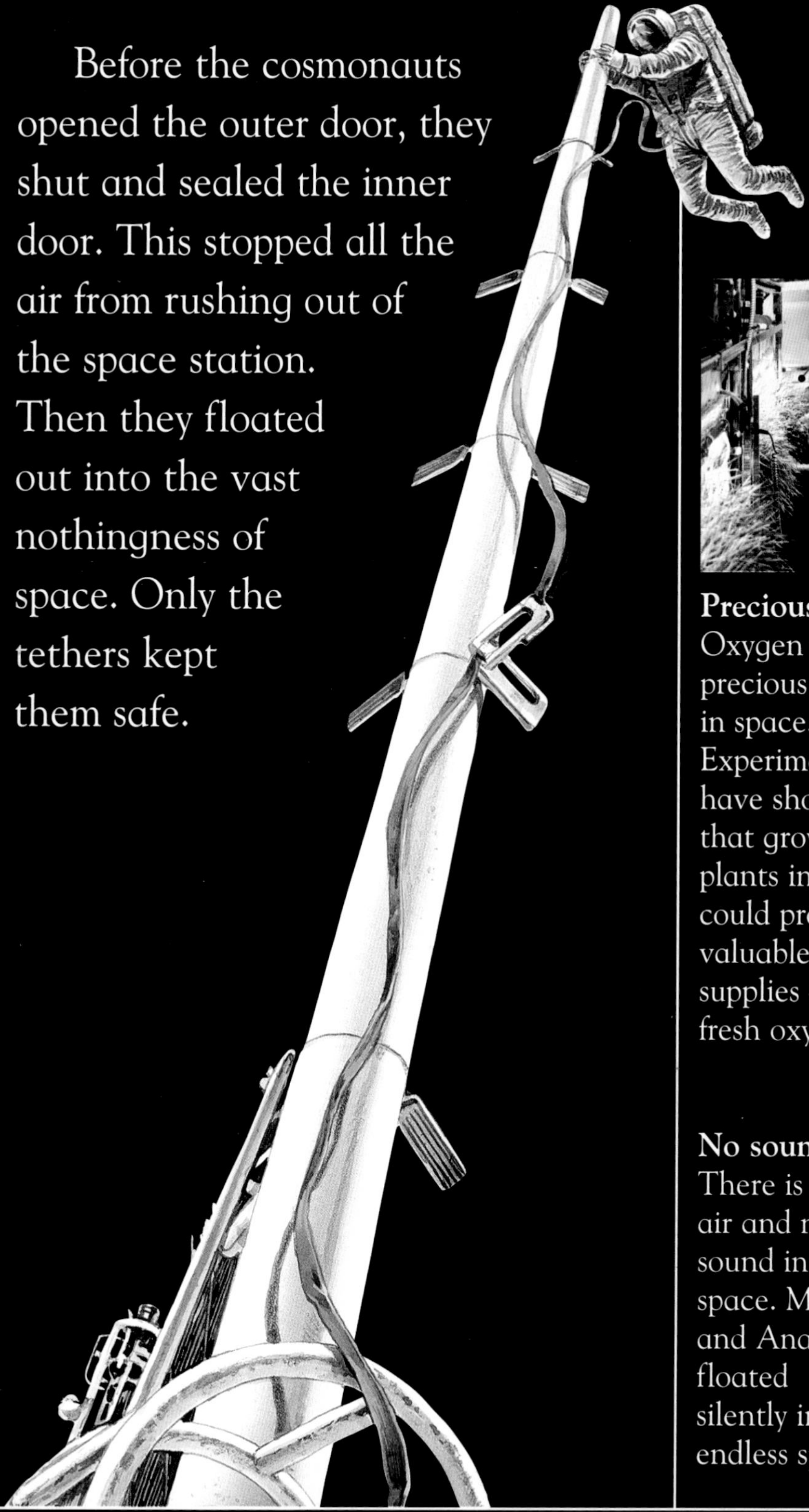

Before the cosmonauts opened the outer door, they shut and sealed the inner door. This stopped all the air from rushing out of the space station. Then they floated out into the vast nothingness of space. Only the tethers kept them safe.

Precious air
Oxygen is precious in space. Experiments have shown that growing plants in space could provide valuable supplies of fresh oxygen.

No sound
There is no air and no sound in space. Mike and Anatoli floated silently in the endless space.

Spacewalk
During their six-hour EVA, Mike and Anatoli orbited Earth nearly four times. It took about 92 minutes for Mir to make one complete orbit of the planet.

Mike and Anatoli moved slowly from Kvant 2 across Mir to Spektr. Mike held the boom tightly and swung Anatoli towards the damaged module. As the hours ticked by, they moved through sunlight and darkness, over and over again. Mike loved the stillness.

Anatoli began to look for the hole in Spektr. He used a knife to cut into the material that covered the module. Pieces of it flew in all directions. Mike watched anxiously. If anything tore Anatoli's spacesuit, he would die. Even a small tear would make the suit useless.

Anatoli tried for hours to find the hole, but he could not find it. He moved the solar panels, though, so that they would collect even more sunlight. He also photographed the damage to Spektr close up. Then the two men re-entered the airlock.

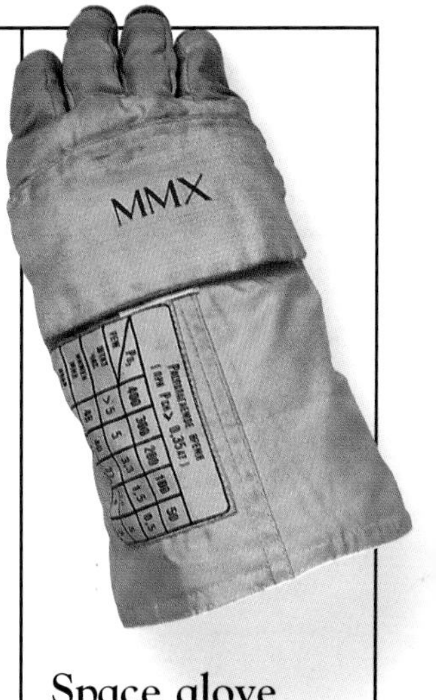

Space glove
During their spacewalk, Mike and Anatoli wore special gloves that left their fingers cold and tingling. Mike described it as being like putting his hand in snow.

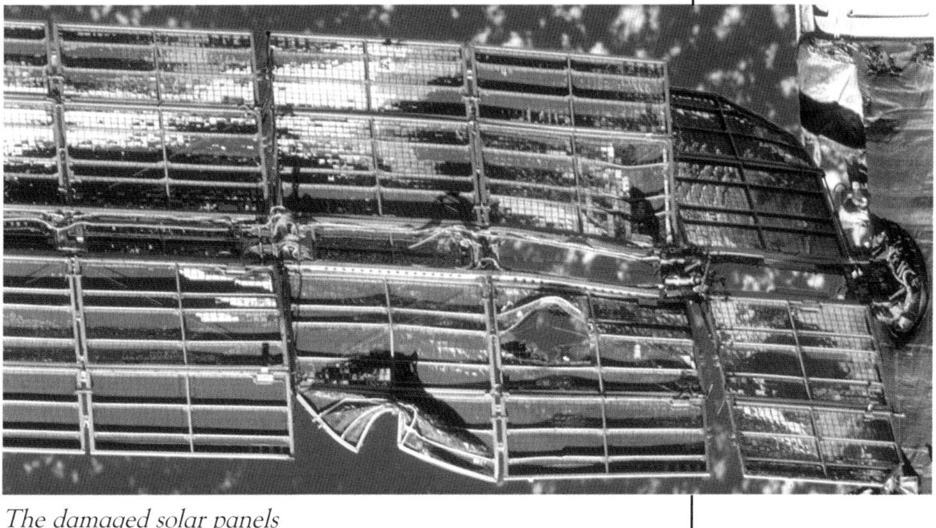

The damaged solar panels

Next in line
Dave Wolf was the astronaut who replaced Mike on board Mir. He was originally going to go to Mir early in 1998, but his mission was brought forwards.

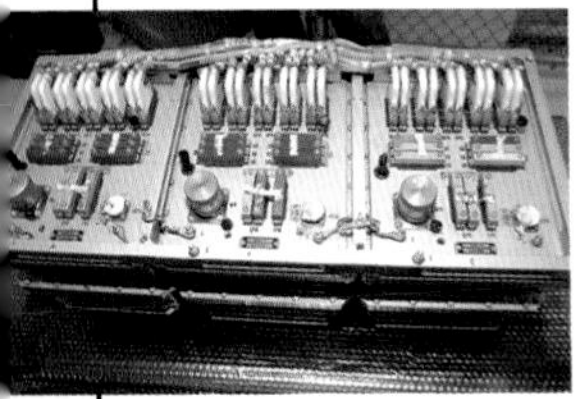

New computer
The Russians asked the Americans to take a new computer to Mir on a space shuttle. It was made and programmed in Russia.

The crew expected life on Mir to return almost to normal, and Mike began to look forward to going home. The space shuttle was due to arrive on 26 September to pick him up and bring another astronaut to take his place.

Then, just two days after the spacewalk, Mir's main computer failed again. Mir spun out of control and the crew had to reposition it.

"Thank goodness it didn't happen during our spacewalk!" said Mike, but he was worried. NASA was wondering whether it was safe to send another American astronaut up to Mir. Mike's return home to Earth might be delayed.

To Mike's relief, *Atlantis* launched on time on 25 September and docked with Mir two days later. When the hatches between Mir and *Atlantis* were opened, the shuttle commander stepped forward, holding a new computer. Anatoli accepted it with a huge smile.

OK for now
Cosmonauts finally installed the new Mir computer on 3 October. After that, the station's computer troubles were over – but only for a while.

Reusable
The shuttle spacecraft are used up to 100 times. *Atlantis* took Mike up to Mir and brought him home again.

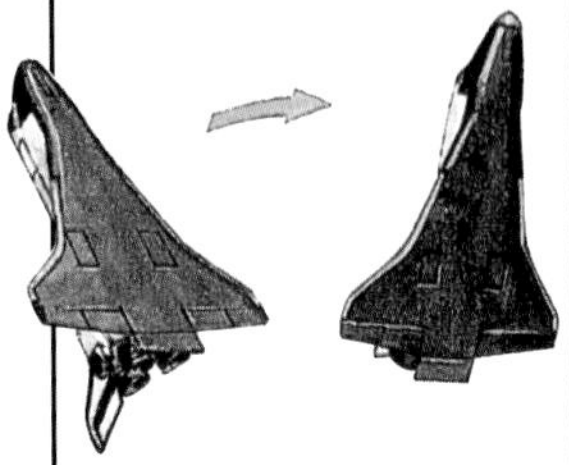

Delay
The wind at Cape Canaveral was so strong on the day planned for Mike's landing that the landing was delayed for a day, until 6 October.

Homeward bound

The hand-over took six days. Then, on Thursday, 3 October, Mike left Mir to travel in *Atlantis* back to Cape Canaveral. Down on Earth, his family was longing to see him again. Jenna and Ian drew a "welcome home" poster.

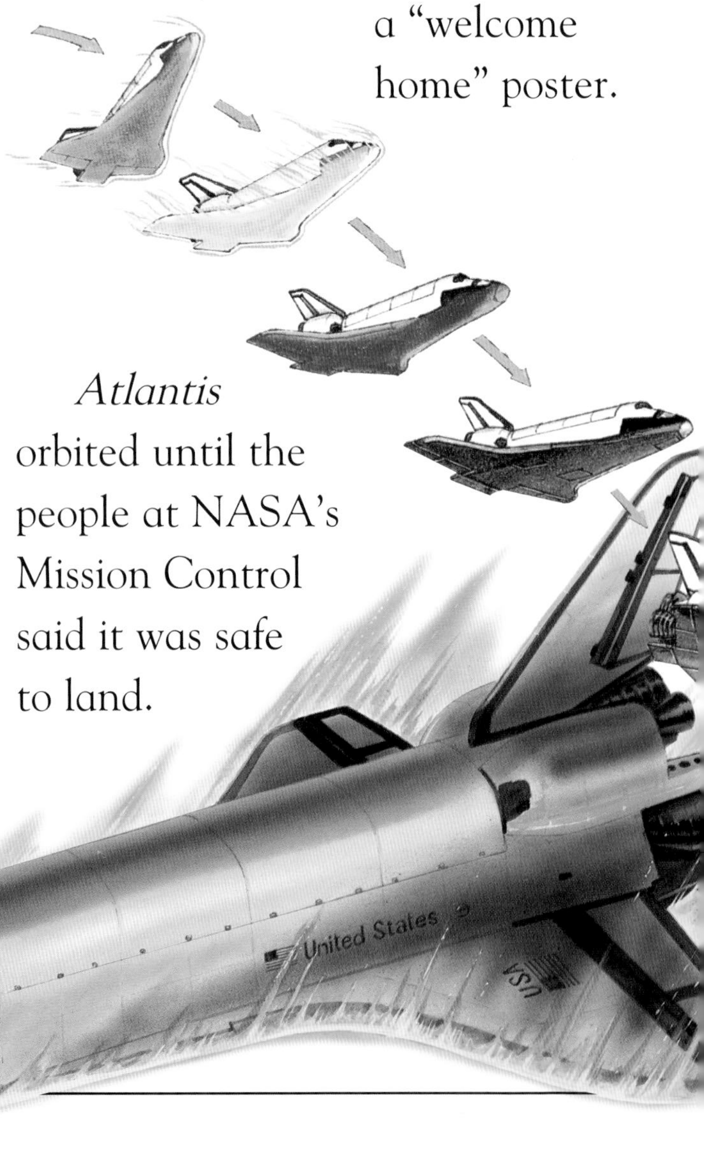

Atlantis orbited until the people at NASA's Mission Control said it was safe to land.

On Sunday, the crew got the go-ahead to land, and Mike's family went to Cape Canaveral to watch. The pilot slowed down *Atlantis* and it began to drop towards Earth. The crew switched off their radios and strapped themselves into their seats for landing.

Atlantis dropped through the thin air high above Earth at about 26,000 km/h. The craft had no engines so it glided down.

Gradually, the air became denser and denser. Friction between the air and the falling spacecraft slowed down *Atlantis* but also made the outside red-hot. A heat shield protected it from the heat and stopped it from burning up.

Radio silence
When the crew switched off their radios, Mission Control lost contact. Then no-one on the ground knew what was happening to *Atlantis*. This always happens between about 85 and 53 km above Earth.

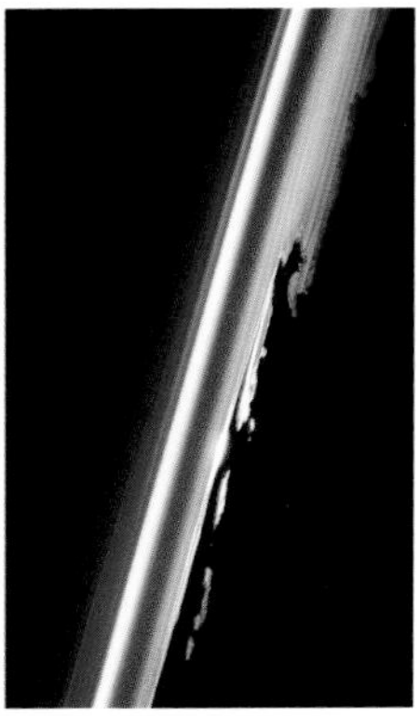

Thin air
The further you go from Earth, the thinner the air gets, until it fades out. This photograph of Earth, taken from space, shows how narrow the band of air is.

Huge glider
Atlantis has wings like an aeroplane's. These have no use in space but help the craft to glide down to land.

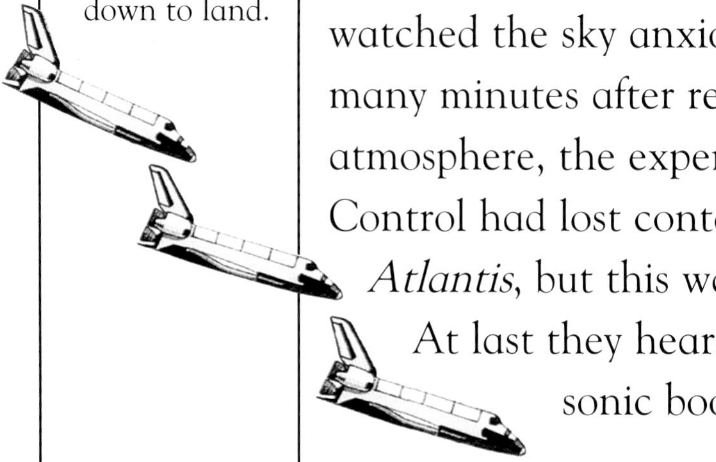

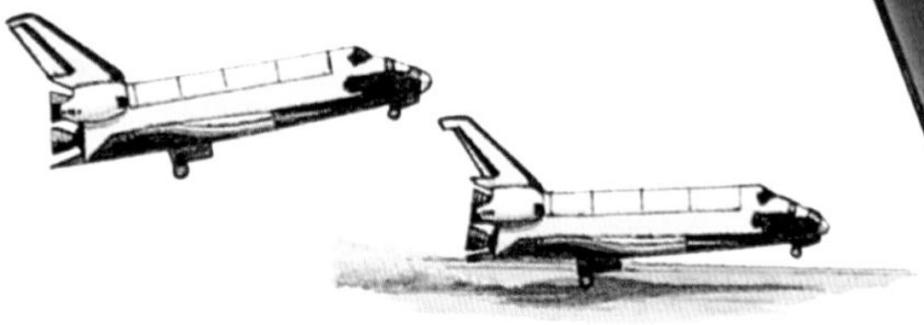

Sonic boom
When an aircraft flies faster than the speed of sound, people below can hear a loud bang called a sonic boom.

The pilot steered *Atlantis* towards the landing site. Soon the spacecraft was gliding no faster than a fighter jet – about 675 km/h.

At Cape Canaveral, everyone watched the sky anxiously. For many minutes after re-entry into the atmosphere, the experts at Mission Control had lost contact with *Atlantis*, but this was normal. At last they heard a dull sonic boom.

Minutes later a tiny speck appeared in the deep blue sky – *Atlantis* was nearly home!

There were just 14 seconds to touchdown. The pilot lowered the wheels and the craft landed on the runway.

Atlantis was still speeding at 410 km/h. A kind of parachute ejected and ballooned out

behind the craft to slow it down. Everyone cheered while Jenna and Ian jumped up and down with excitement.

"Welcome home. Excellent job," said Mission Control as *Atlantis* came to a halt.

Drag chute
The parachute that ejects behind the spacecraft is called a drag chute. It acts as a brake by dragging through the air.

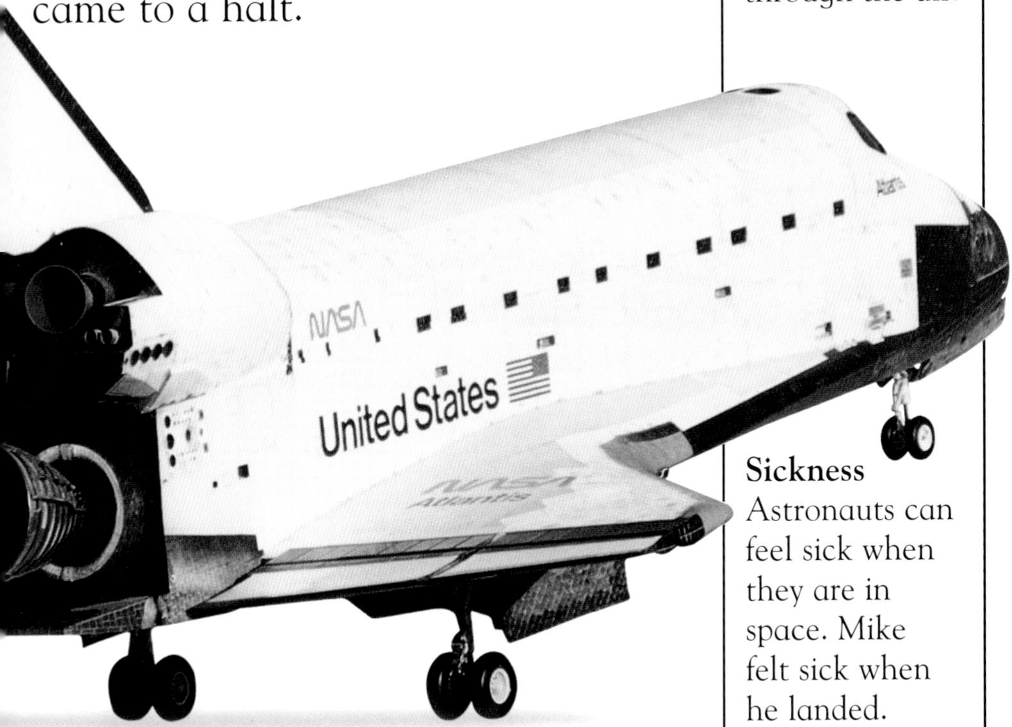

Sickness
Astronauts can feel sick when they are in space. Mike felt sick when he landed.

Work out
Weightlessness in space can weaken muscles because moving around takes less effort. Astronauts must exercise for two hours a day to keep fit.

JFK
In 1960, US President John F. Kennedy encouraged astronauts to go to the Moon. The Kennedy Space Centre in the USA is named after him.

The astronauts waited for about an hour for the shuttle to cool down. Then they came out. Mike found it difficult to walk. Although he had exercised on Mir, his muscles had become weak. While he was weightless in space, he had not had to work as hard as he did on Earth. But nothing stopped him from hugging and kissing his family. "It's great to be back," he said, and Jenna and Ian agreed.

Mike spent that night with his family in rooms in the Kennedy Space Centre.

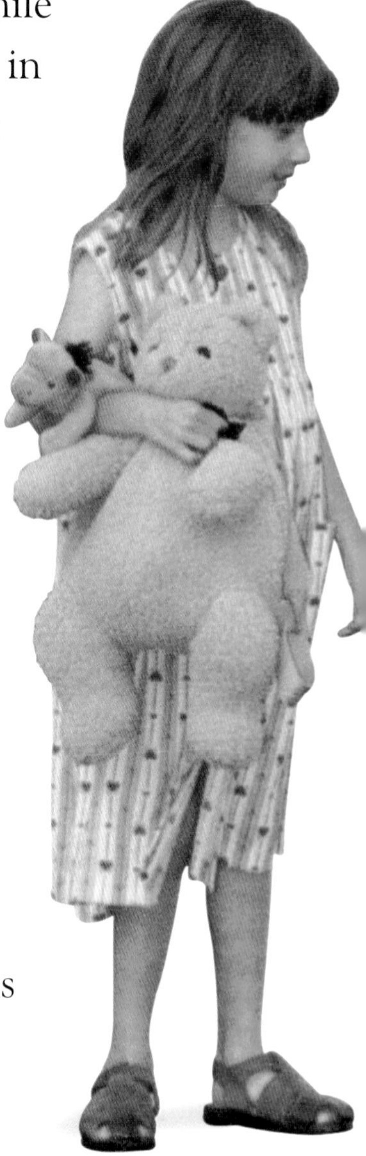

He was asked what he would like to eat for his first meal back on Earth. "Lasagne and pizza, plus a lot of beer and chocolate," he replied.

Later, Mike said that being on Mir was the most difficult thing he had ever done, but he also remembered what President John F. Kennedy had once said: "We do not attempt things because they are easy, but because they are hard, and in that way we achieve greatness."

New strength
It took several weeks for Mike's muscles to get strong again.

The story continues

Mir did not break down because it was badly built. In fact, it survived the crash with Progress only because it had been so well designed.

The Russians and the Americans wanted to keep Mir going. NASA paid for their astronauts to live on Mir to learn what it was like to spend a long time in space. Russia needed the money to fund their own space programme. The astronauts wanted Mir to keep going too. A new space station was being planned and they wanted to get as much experience of long missions as possible.

The year after Michael Foale returned from Mir, American space shuttles and Russian rockets began to take the parts of the new space station into space.

Next mission
Crews were chosen for a space shuttle mission to Mir about two years in advance. This was the crew picked to fly in September 1997, including Mike Foale, who flew home with the crew.

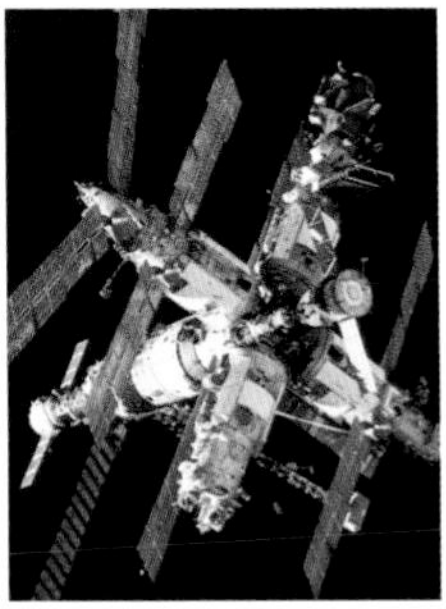

Goodbye, Mir
Mir was left unmanned in 1999. In 2000, a Russian crew went up to see if Mir could be put back in use.

The new space station is being launched into orbit around Earth, piece by piece. Europe, Japan and Canada are helping to build it. This artist's impression shows what it will look like when it is completed.

Good advice
Mike was able to use his experiences on Mir to advise the people who are building the new space station.

Solar panels
Modules for laboratories and living quarters
Zarya control module was the first section to be put into orbit

Mike Foale hovers above Discovery's *cargo bay during a spacewalk to repair the Hubble Space Telescope.*

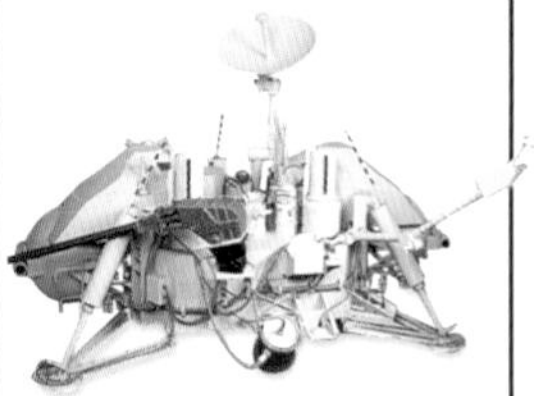

Space probes
Robotic space probes have visited all the planets in the solar system, except Pluto.

While the new space station is being built, spacecraft have continued to explore space. In 1997, an unmanned spacecraft landed on Mars, 56 million km from Earth. A spacecraft was also launched that will reach Saturn in 2004.

In December 1999, Mike Foale went up in the space shuttle *Discovery*, this time to repair the Hubble telescope. This huge telescope is in orbit more than 600 km above Earth. It takes clear

pictures of distant stars and galaxies. By 1999 it had been in space for nearly ten years. Parts of it were wearing out and had to be replaced. The astronauts made three spacewalks on this mission to make the repairs.

There will be many more space adventures. NASA may send astronauts to Mars, and there are plans to build a space station on the Moon. One day there may even be a hotel orbiting Earth. Who knows?

Long journey
It took more than six months for the spacecraft carrying the Viking lander to reach Mars. Viking found no signs of life on the planet.

Gyroscopes
Astronauts Steven Smith and John Grunsfeld made repairs to the gyroscopes that kept Hubble pointing in the right direction.

Glossary

Air pressure
The force of air. The more air the greater the air pressure.

Airlock
A hatch with two doors from which air can be sucked out and pumped in.

Astronaut
An American person who travels into space.

Atlantis
The spacecraft that carried Michael Foale to and from Mir.

Battery
A device used to store electricity.

Boom
A long movable metal arm like a fishing rod.

Charged
A battery is said to be charged when it is full of electricity.

Cosmonaut
The Russian word for an astronaut.

Countdown
A counted check down to zero, when a certain event, such as the lift-off of the shuttle, is planned to take place.

Docking
When two spacecraft, such as *Atlantis* and Mir, lock together in space.

EVA
Extravehicular activity, also called a spacewalk, when astronauts work outside their spacecraft in space.

Gravity
The force that pulls things towards Earth and gives them their weight.

Ground Control
The scientists and computers on Earth that control a space mission.

Hatch
A type of door. In space, hatches can be made airtight.

Heat shield
Tiles on a spacecraft that protect it from burning up as it falls through the air.

Mir
The space station in which Michael Foale lived for four months.

Module
A section of a space station or spacecraft.

Node
A module on Mir to which four other modules were joined, and where Soyuz was docked.

Oxygen
A chemical in the air that all living things need to breathe to stay alive.

Progress
An unmanned Russian spacecraft that brought supplies to Mir and took away rubbish.

Remote control
Controlling something by computer from a distance.

Rocket boosters
Rocket engines that give the space shuttle power to launch into space.

Solar panel
A device that changes the energy in sunlight into electricity.

Soyuz
The name of the Russian spacecraft that took cosmonauts to and from Mir and acted as the emergency escape craft.

Space shuttle
American spacecraft that are launched from Earth by rocket and return like gliders to land on a runway. They can be used again and again.

Spacesuit
Protective clothing worn by astronauts. Spacesuits that provide air, water and warmth are worn during EVAs.

Spektr
The module on Mir that was hit in the collision with a Progress spacecraft.